Happy Cells!
by
Bronwyn H. Tollefson

Illustrated by

Haude Levesque, PhD

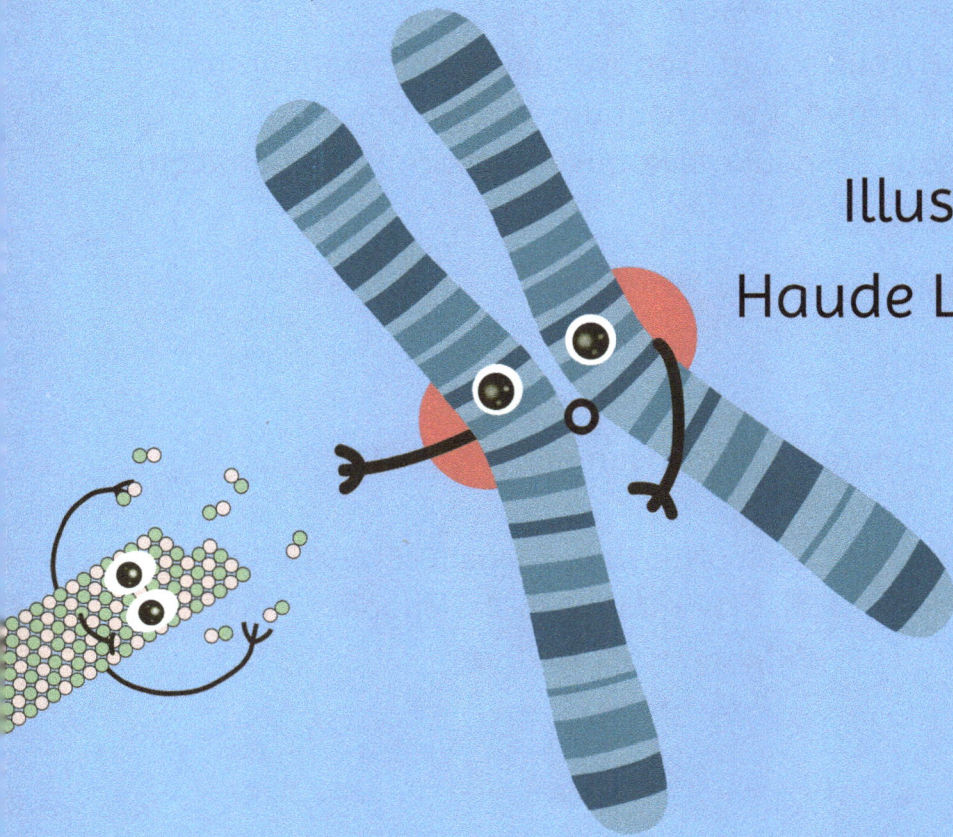

For more information about this book,
contact Bronwyn H. Tollefson at: Happybookinquiries@gmail.com
For more information about the illustrations,
contact Haude Levesque at: haude.levesque@science-illustrations.com

Library of Congress Control Number: 2020907131

ISBN: 978-1-7348886-4-5 (hardcover) ISBN: 978-1-7348886-3-8/978-1-7348886-1-
4 (paperback) ISBN: 978-1-7348886-2-1(ebook)
First edition August 2020

To all who have inspired me along the way.
B.H.T.

To my curious boys, Alexandre and Aurelien.
H.L.

3

Hello there! I am excited you decided to open my book! I wanted to provide you with a brief introduction to the science behind this story before you begin reading. This book is all about a process called mitosis (pronounced like my-toe-sis), which is present in humans, animals, and all living organisms! In fact, it is happening right now inside everyone! Where this process takes place is in teeny-tiny structures in our body, called cells. These cells multiply, grow, and connect to each other to create our heart, skin, nose, and so much more! To keep our body working strong, they need to continue dividing and multiplying throughout our life.

Inside these cells, there are even smaller structures that move and are very active, and these are called organelles. Overall, organelles and other cell structures follow a specific set of instructions during mitosis to help cells duplicate its information (known as DNA) and continue growing and dividing:

Step 1: Copy all the DNA inside the cell. Now you have two identical copies!

Step 2: Keep growing and growing!

Step 3: Then, wrap the DNA tightly into an organized bundle (called a chromosome- there are several of them inside one cell) and attach both identical chromosome copies at their center.

Step 4:   Attach long strings of fiber to each of the centers of every chromosome pair and line all of them, from head to toe, in the middle of the cell.

Step 5: Pull on each side of the fibers to separate identical chromosomes from each other!

Step 6: Stretch the whole cell, separating even more chromosomes and organelles, so much that it is so close to splitting in half.

Step 7: Now, completely divide the cell in half, so you now have two identical cells.

Step 8: Repeat!

There are quite a lot of components and activity involved in mitosis overall. In fact, scientists are still learning even more about it every day! As you read through this story, you may come across words that are in bold. These bolded words highlight very important structures and processes in the order that they appear in the story. If you want to know more about the names in bold, look in the glossary at the end of the book (page 32) to find out cool information about them!

Happy **cells**, happy cells

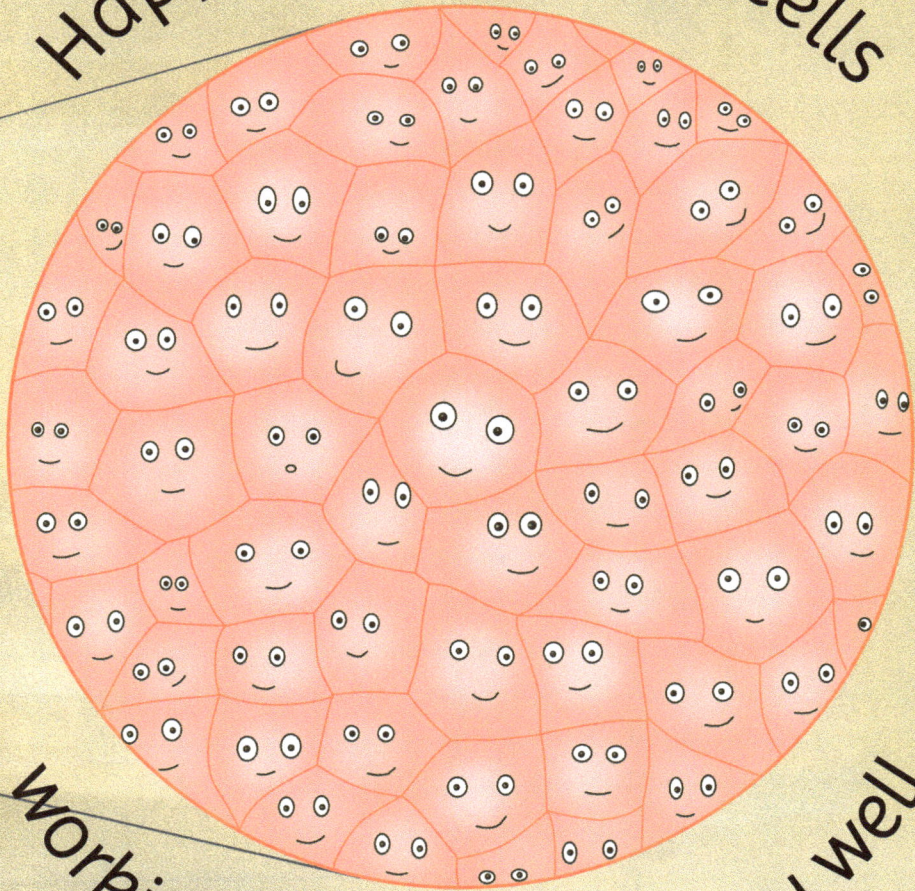

working hard to keep you well.

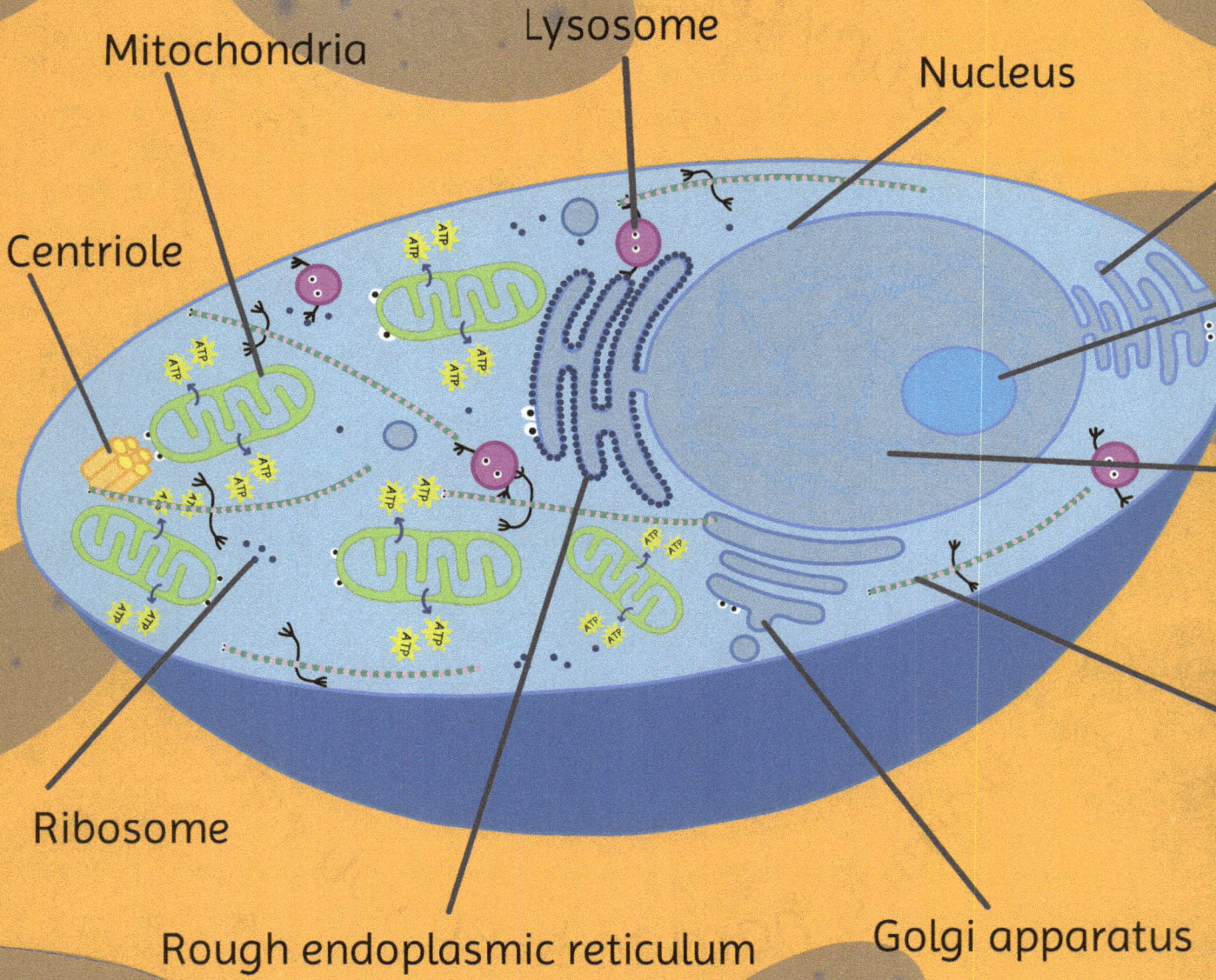

Mitochondria

Lysosome

Nucleus

Centriole

Ribosome

Rough endoplasmic reticulum

Golgi apparatus

ATP

8

Endoplasmic reticulum

Nucleolus

Happy cells have many

organelles. Each one has a

special role.

DNA

They work together everyday

to make you, you! As a whole!

Microtubule

These cells have
important friends
and together, they make
the cell factory succeed!

They work for hours and hours, day and night,
and they do it with a tremendous amount of might!

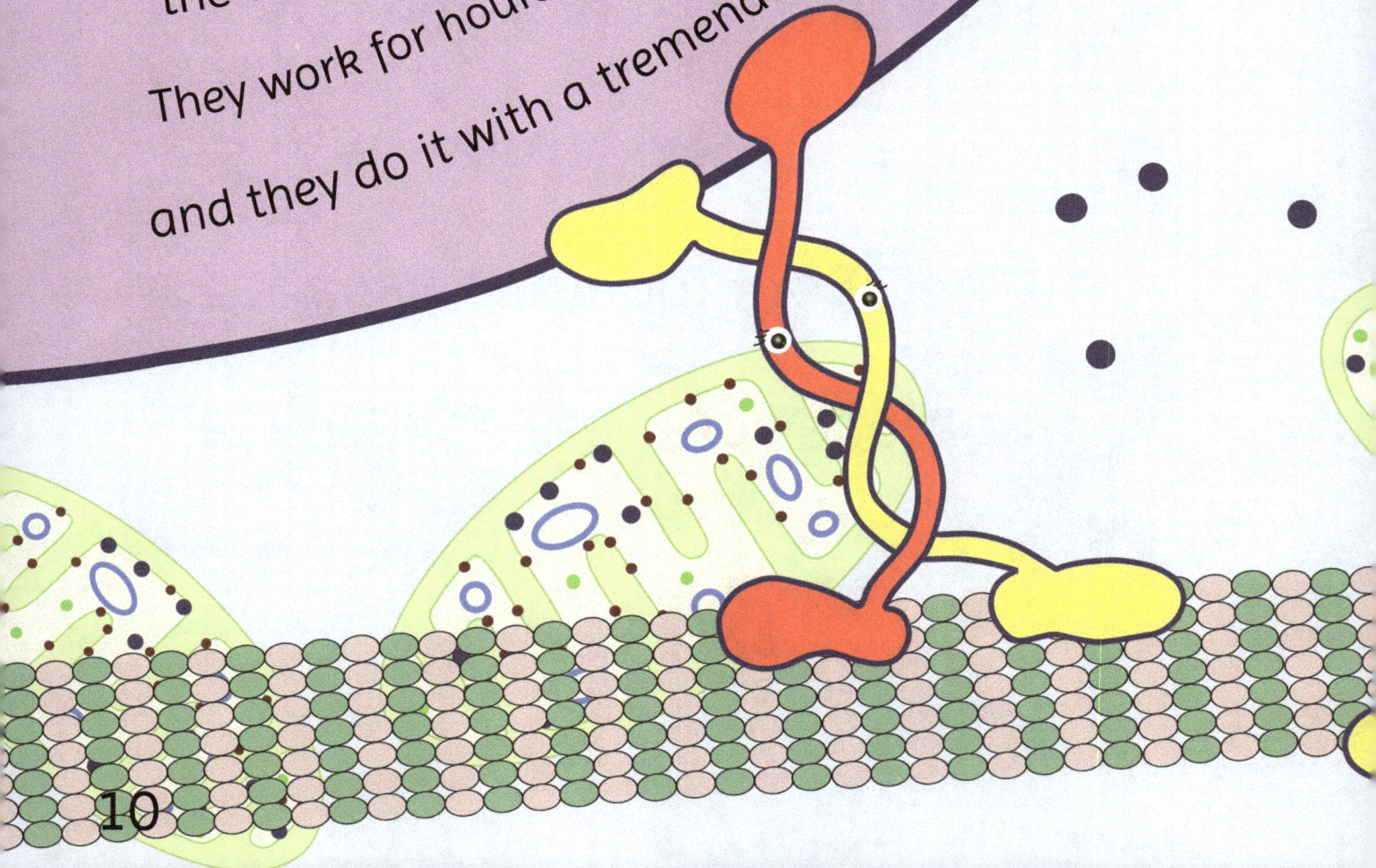

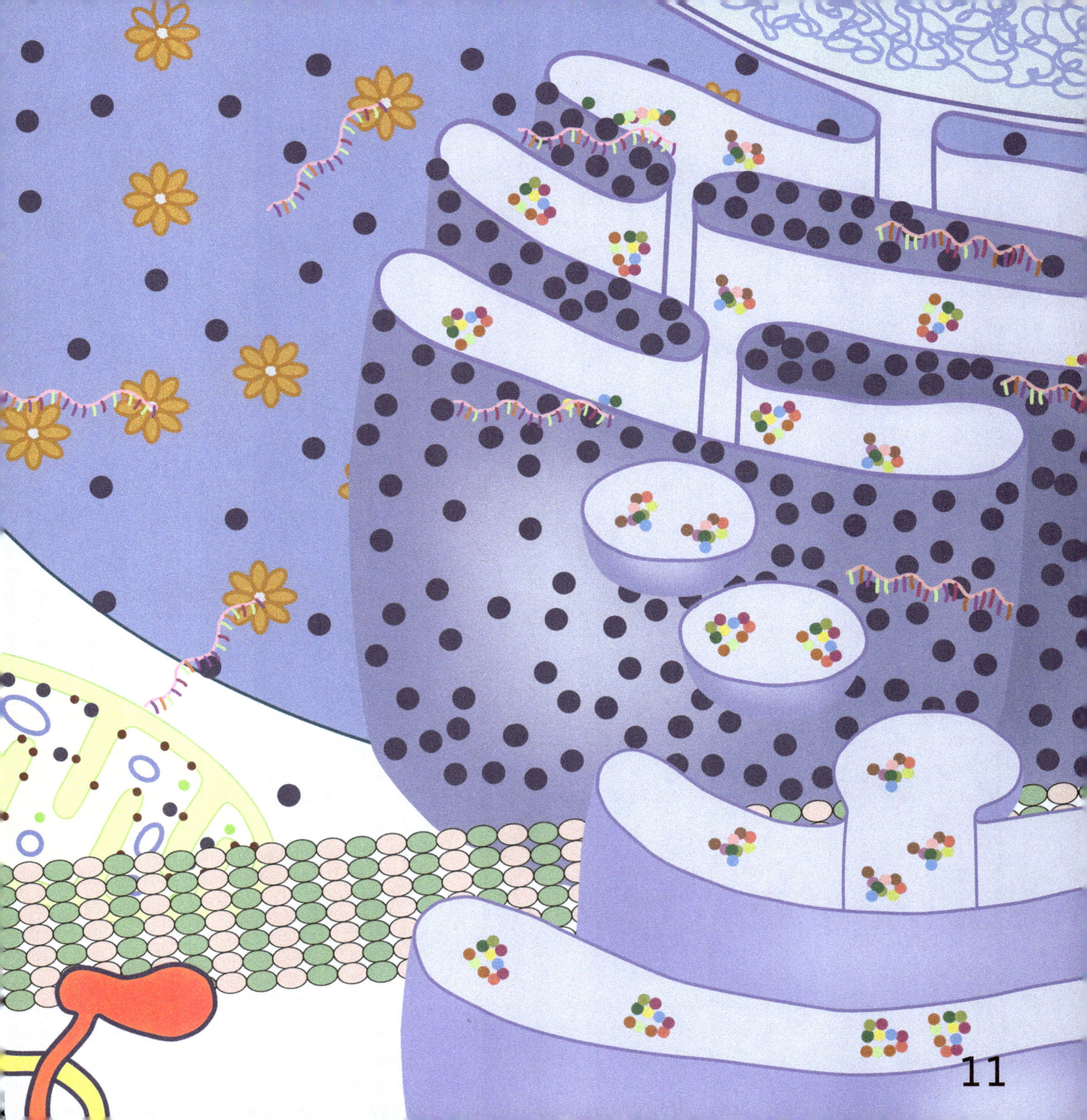

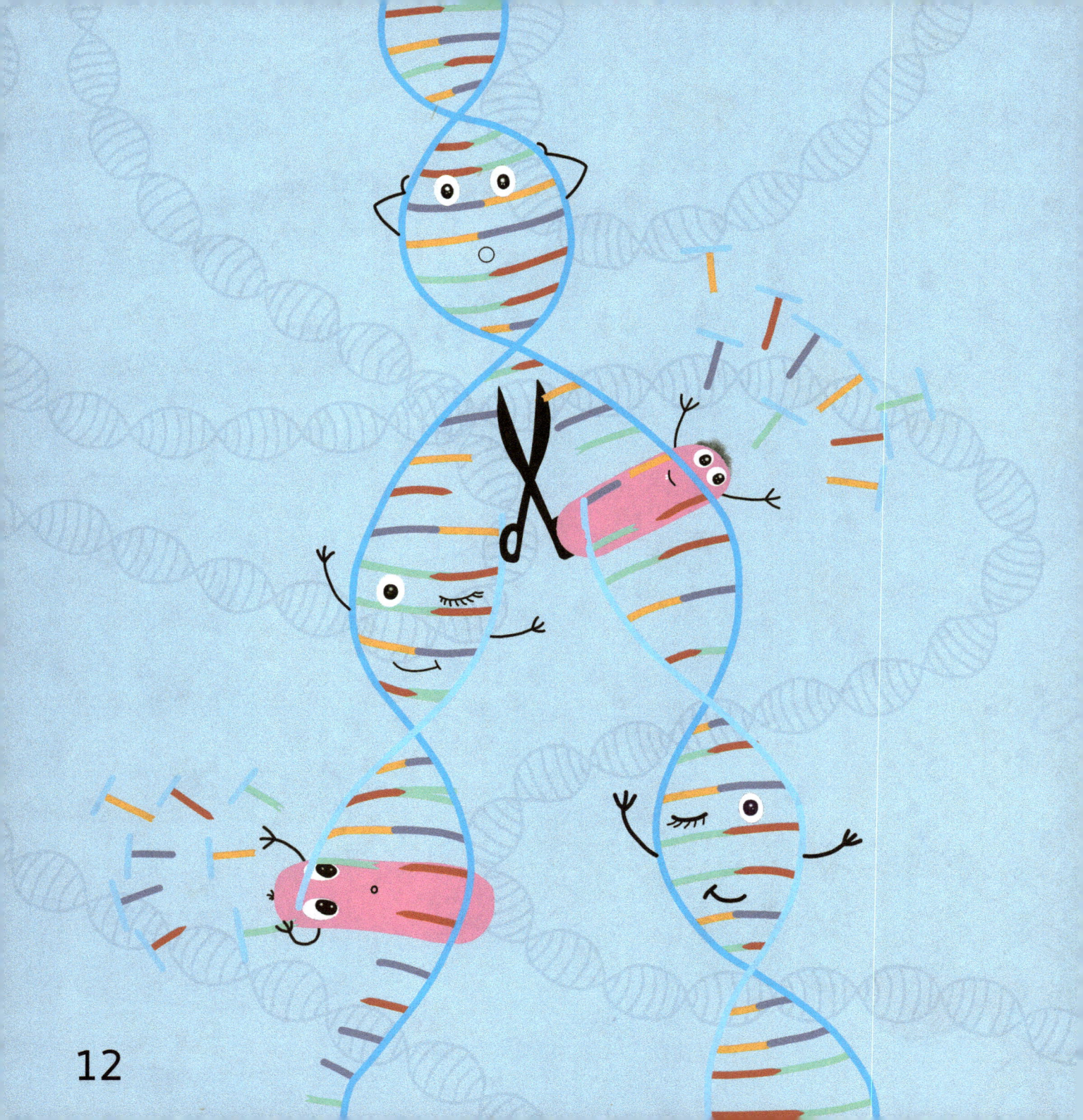

The first friend, Danny **DNA**, copies himself and makes sure everything is the same.

While **DNA Polymerase** makes the new DNA bind, his friend, **Helicase**, helps him unwind.

Look! The cell factory grew! There was one DNA friend, but now there are two!

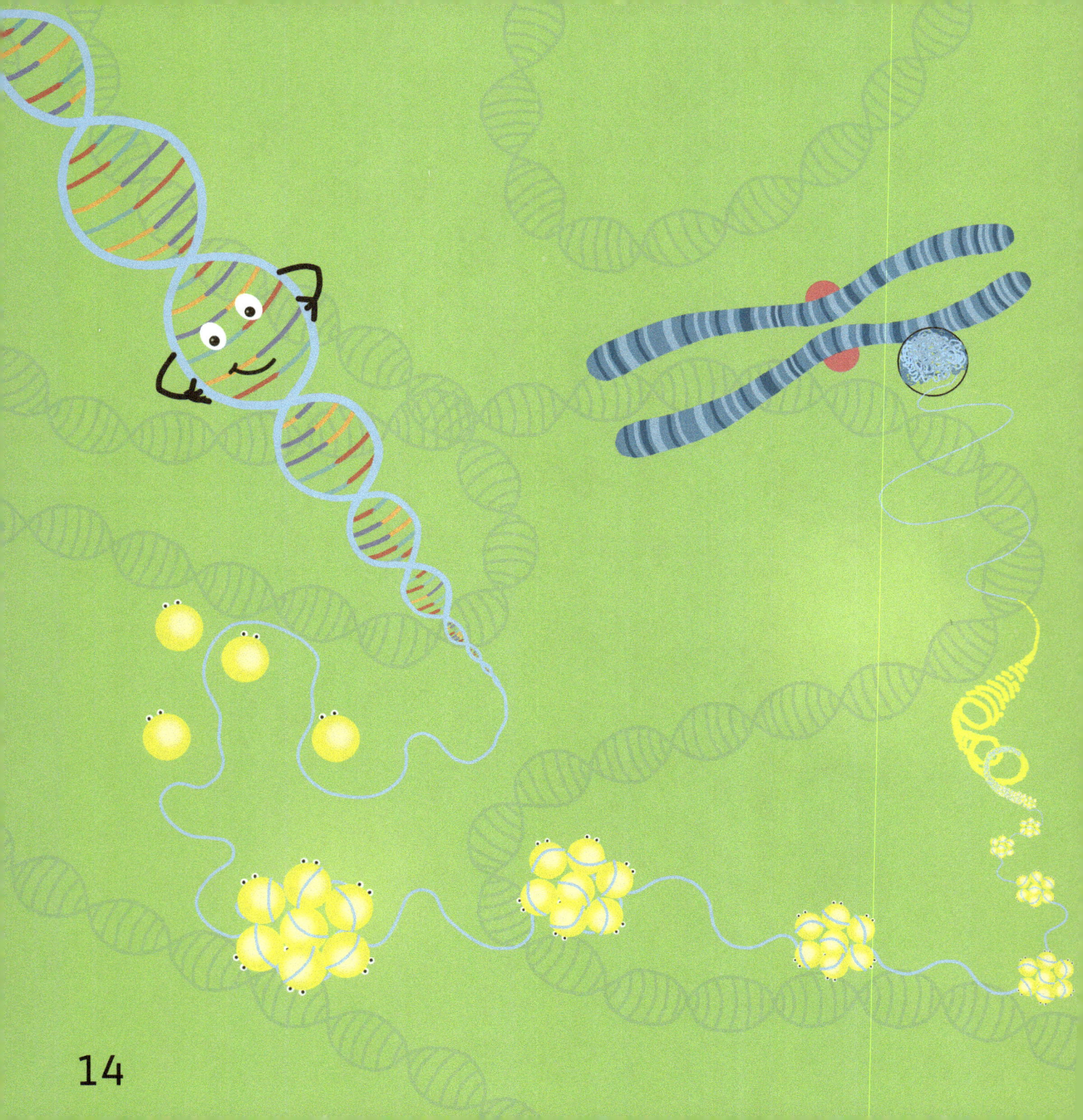

Then Danny DNA decides to spin.

He swirls and twirls and gets tightly packed in.

Henry **Histone** helps Danny DNA roll.

They are now all organized, and that is the goal!

From this, a new friend arrives.

His name is Chris Chromosome!

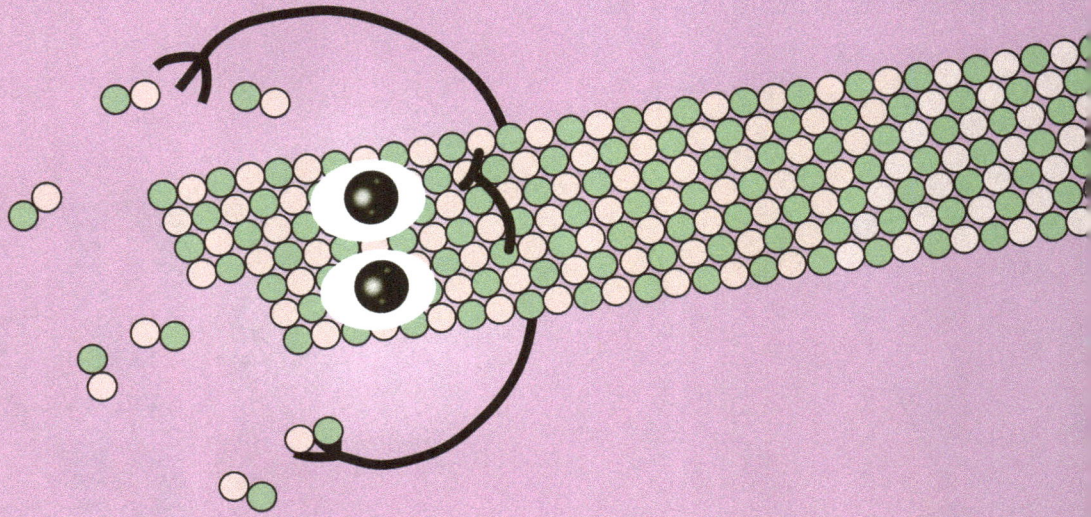

A new organelle, Sammy **Spindle**,

thinks her friends are just grand.

She steps right up and gives Danny

DNA a hand.

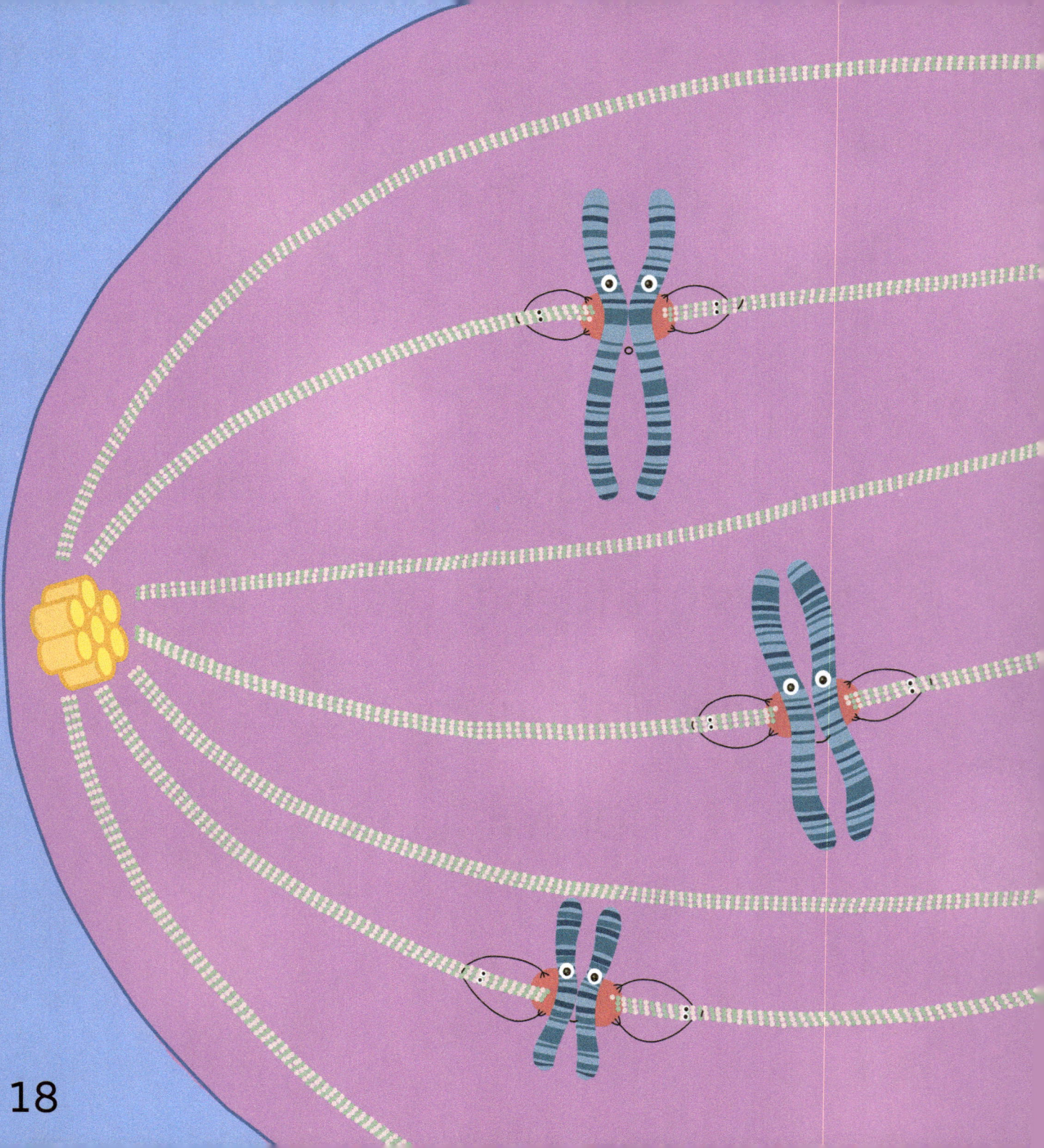

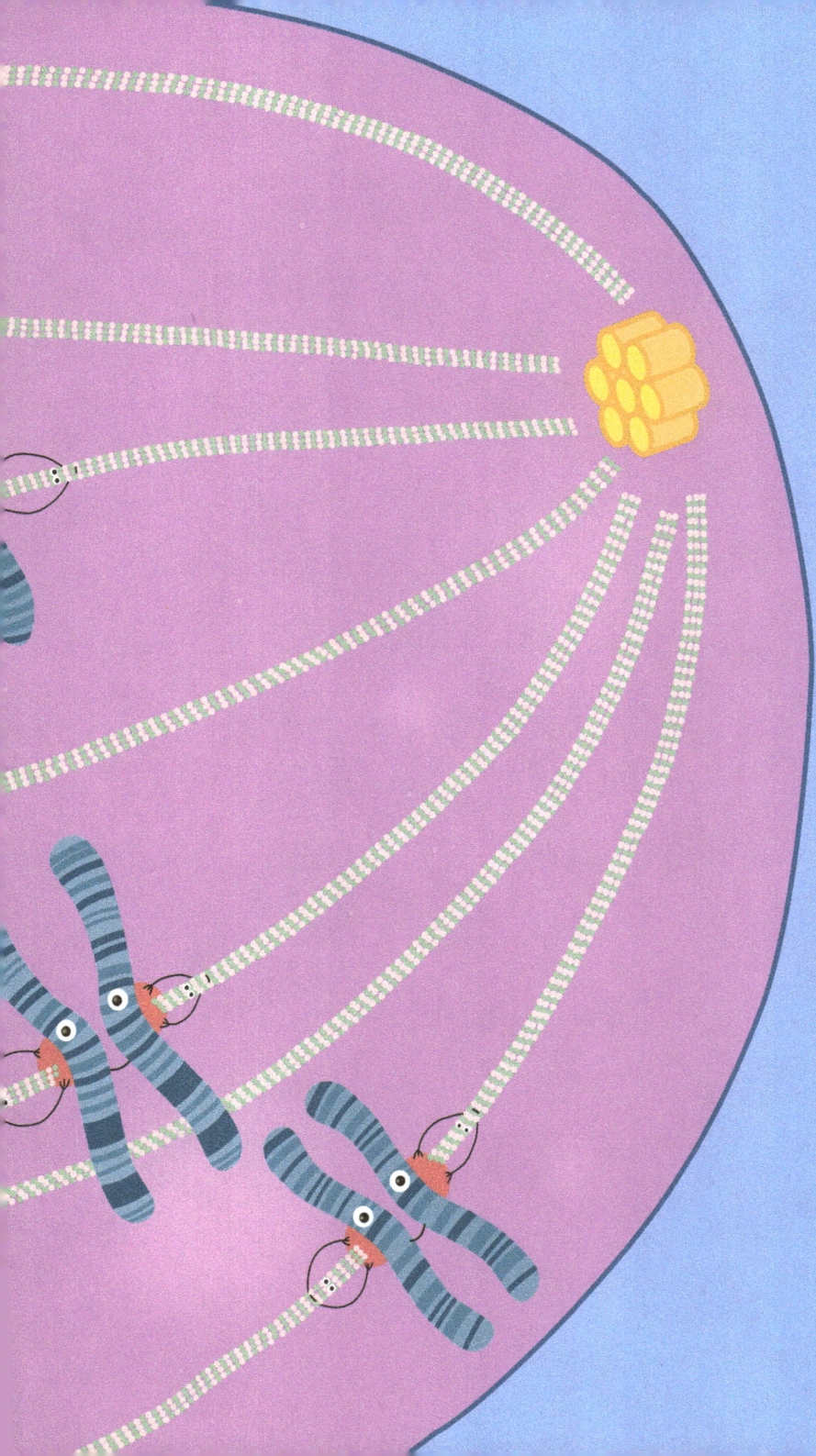

Then Sammy Spindle says,

"Let's flee and go to the middle!

There's so much to see!"

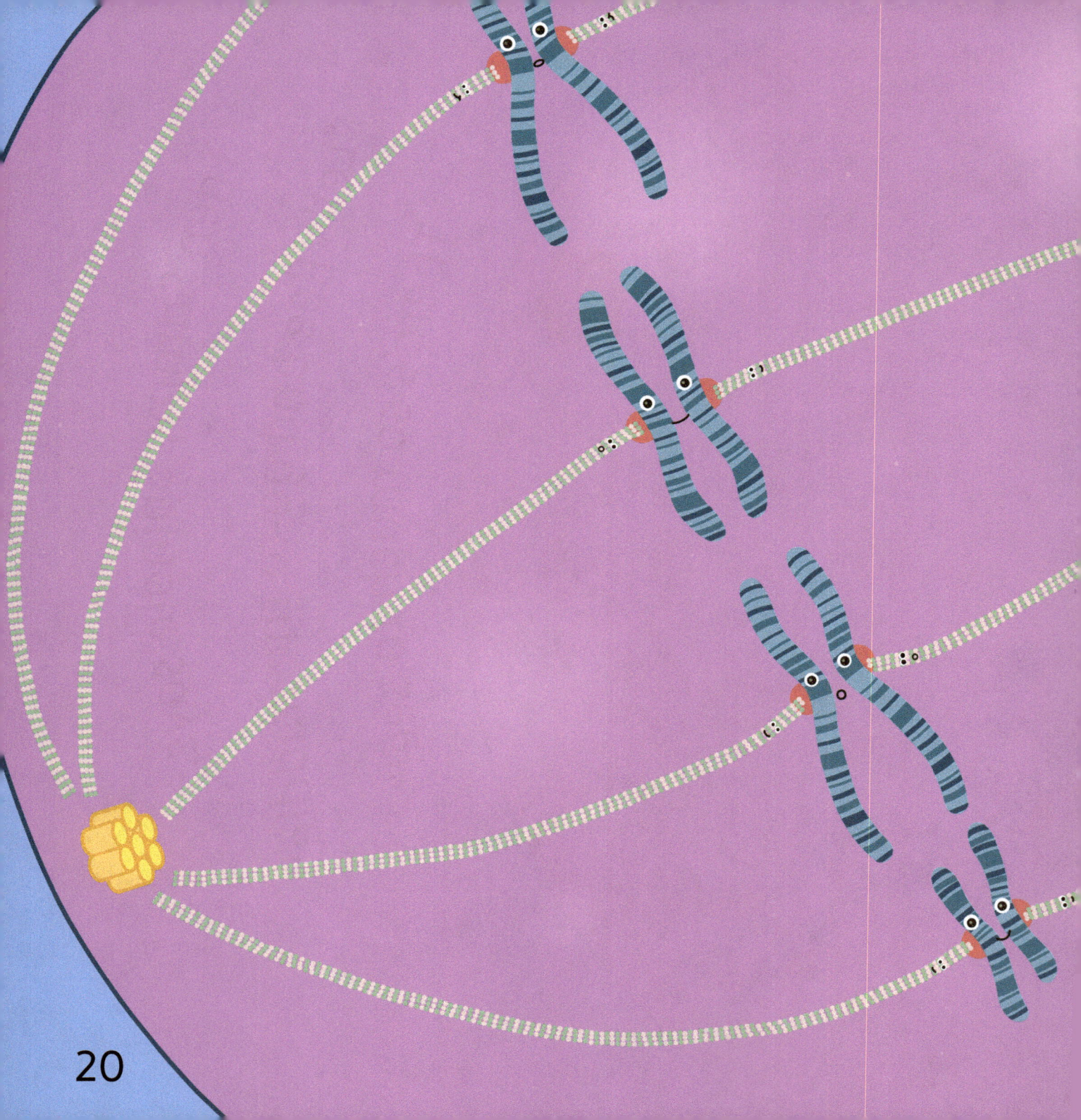

20

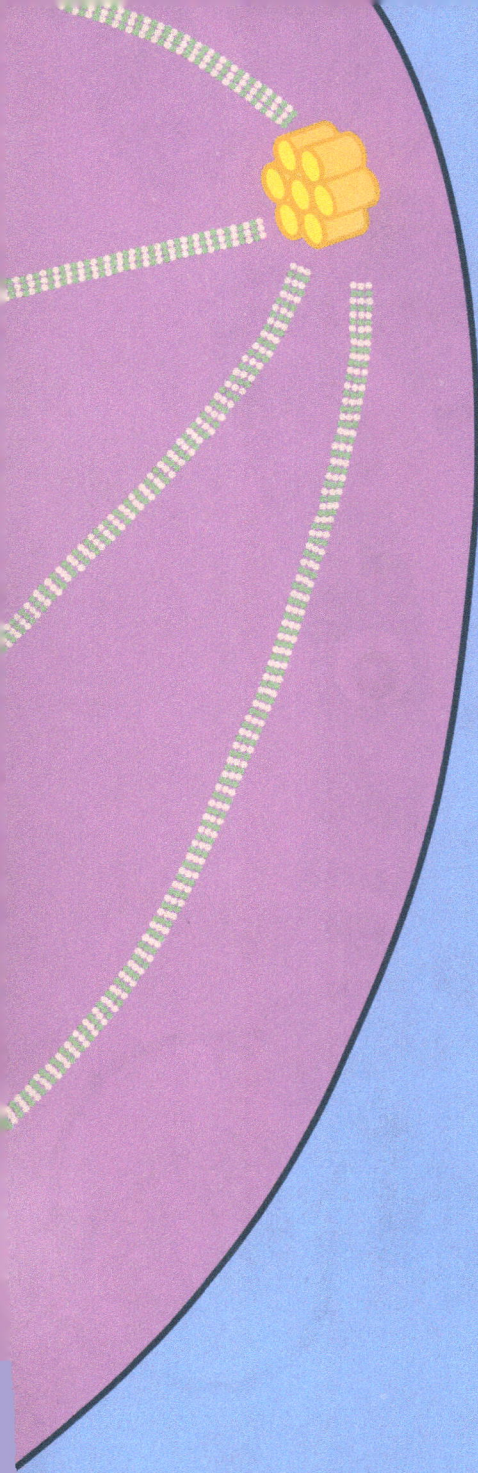

After a while, each Danny DNA gets tired and decides to go, but there is only one problem: alone they'd be too slow. Oh no!

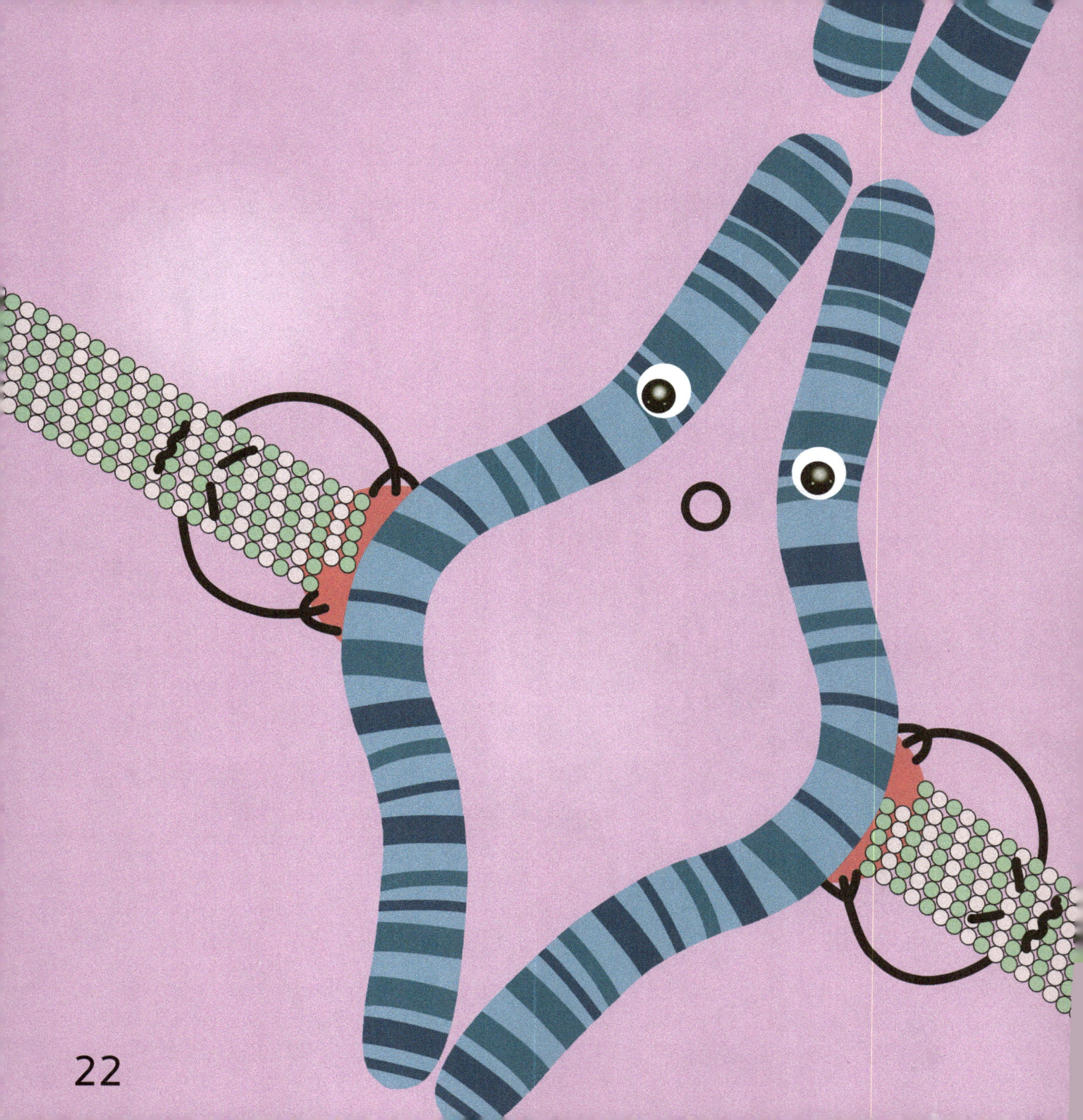

22

In comes Sammy Spindle. She's

got nothing to hide!

With a great big tug,

she helps each Danny DNA get to

his own side!

23

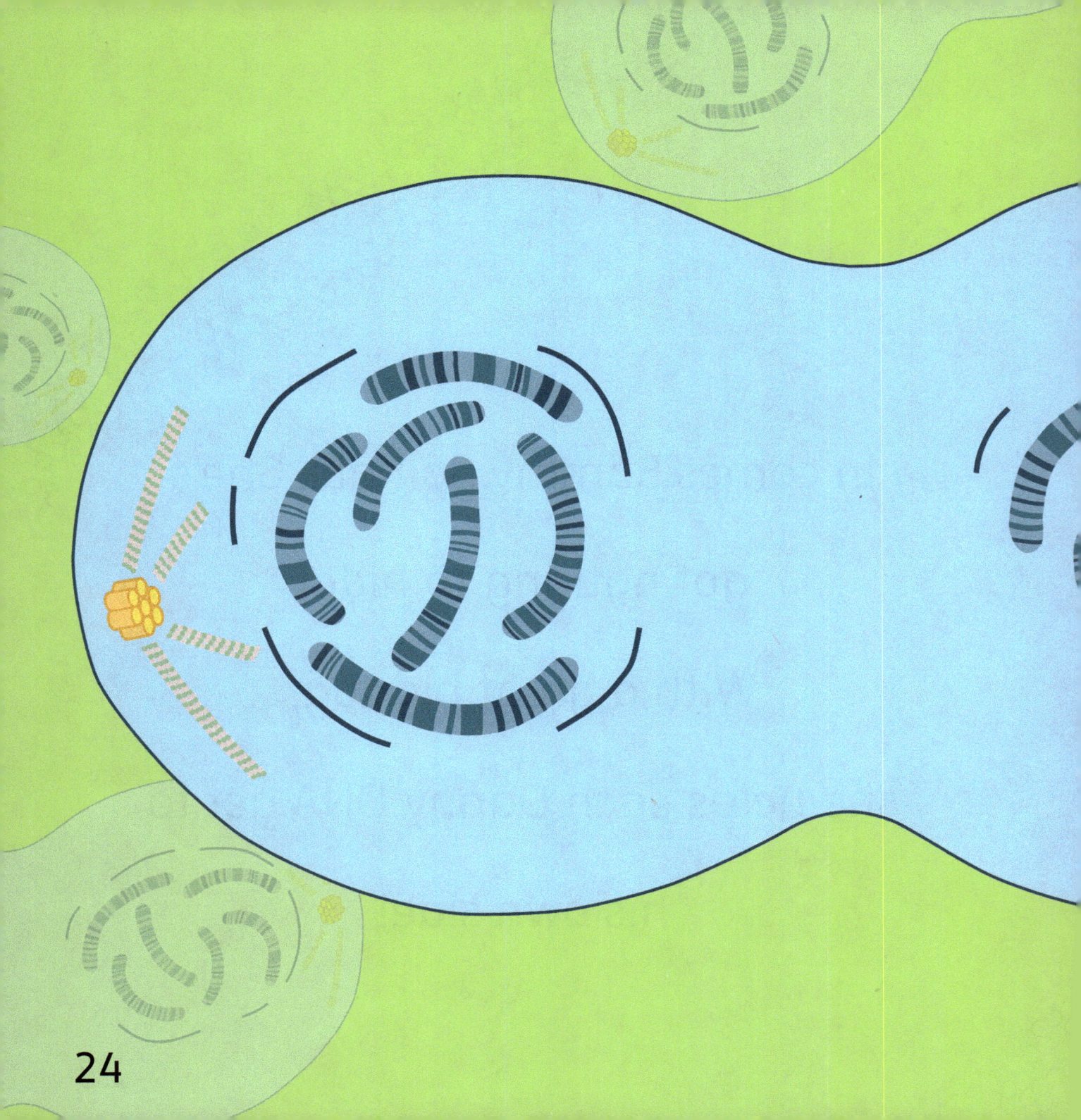

At each side, there are even more friends,

and the DNA become crowded!

There isn't much room.

There isn't much space,

so the cell factory

decides to split...

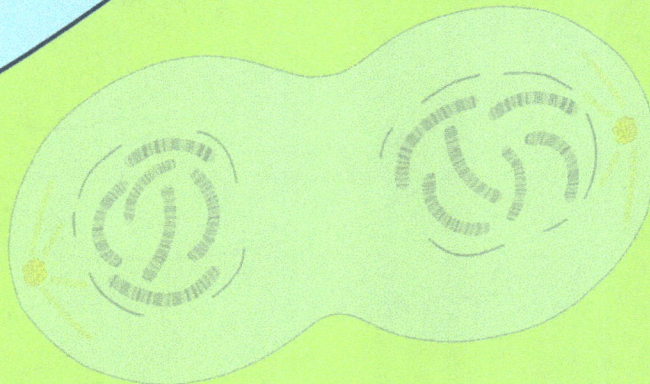

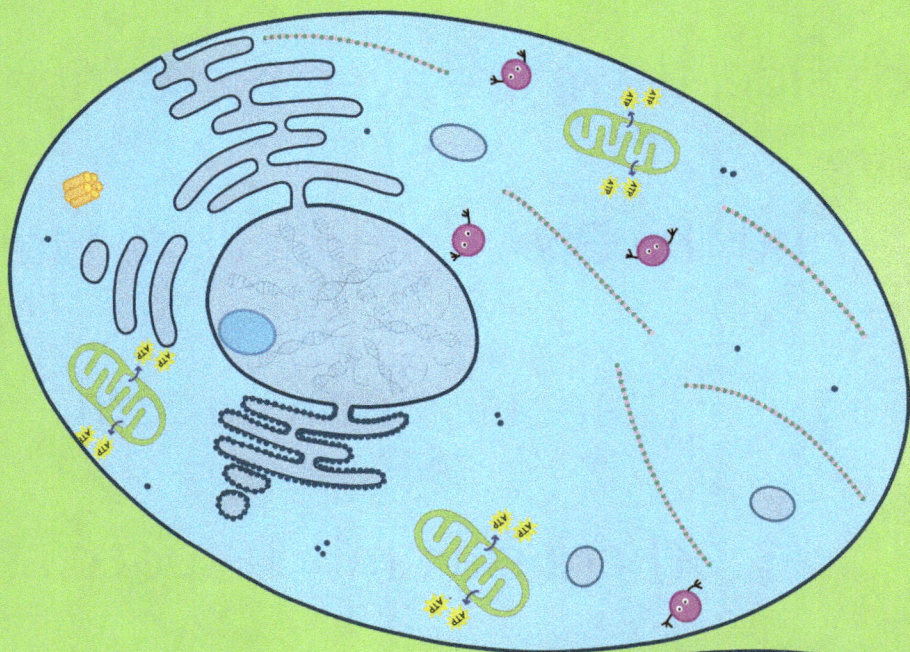
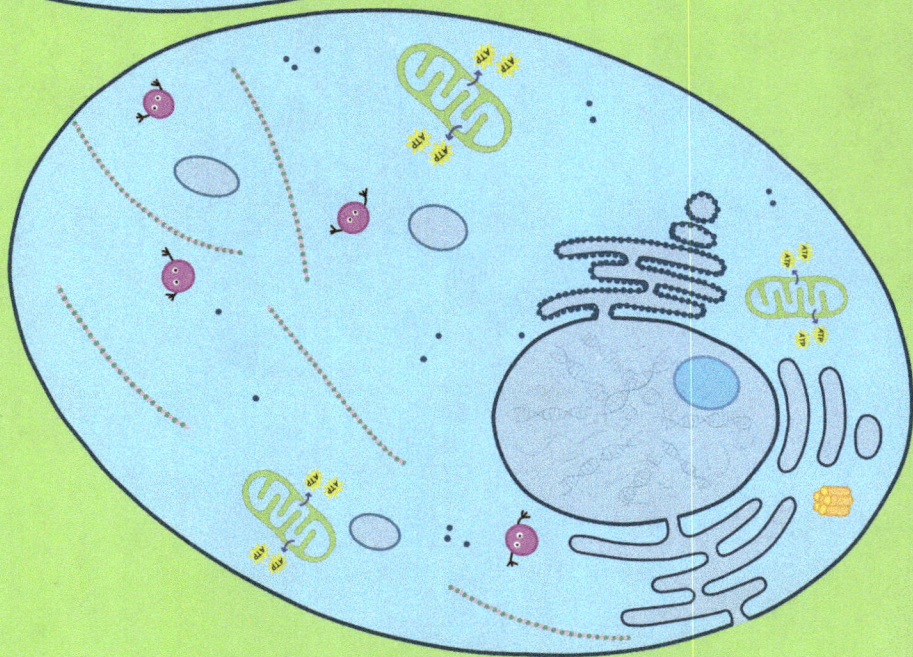

One happy cell turns to two happy cells-

each with great friends, but just then,

the organelles say, "Let's do it again!"

Happy cells, happy cells

working hard to keep you well.

They are important to me and important to you!

It's true!

Because happy organelles make happy cells,

and happy cells make...

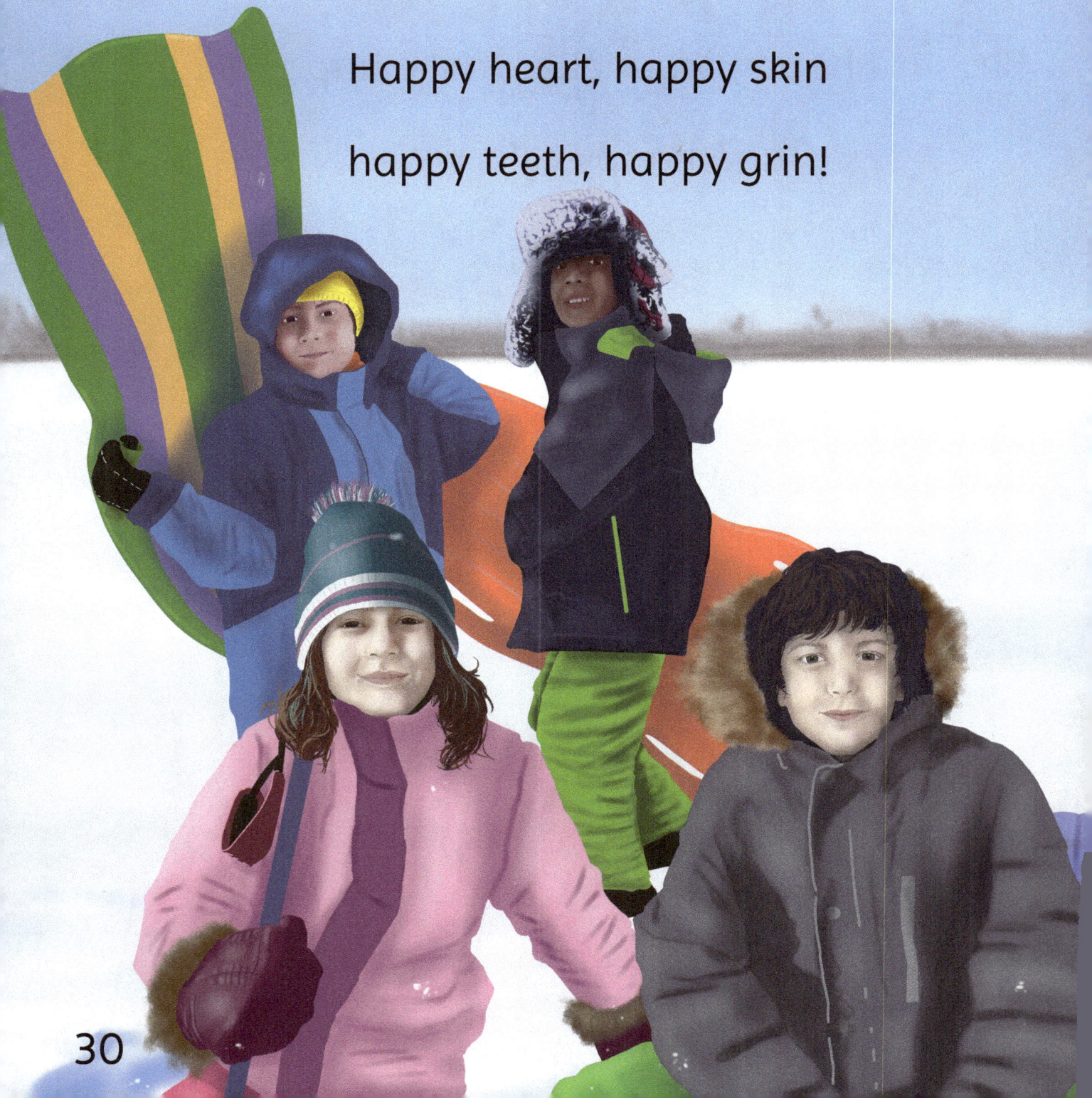

Happy heart, happy skin

happy teeth, happy grin!

Happy tummy, happy eyes

happy bones...

happy lives!

Glossary:

Cell: A small, living structure that keeps splitting in half to make muscles, bones, eyes, and more!

Organelle: Tiny little "organs" that live inside of cells. They help cells live and split in half when they are ready.

DNA: A very important piece of information that is in every single cell in your body. It has all the "information" that your body needs to work every day. When cells split in half, it needs to copy this information!

Helicase: A small structure that helps untwist and separate DNA so that the DNA can copy itself easier.

DNA Polymerase: A structure that makes new DNA from the old DNA. It's like taking a Lego from a huge pile to create an airplane. You look for the exact Lego that you need and once you find it, you attach it to your airplane!

Histone: A tiny "ball" that attaches to DNA to help it pack into an organized chromosome.

Chromosome: When DNA gets tightly wrapped together, a chromosome is formed. This makes it easy for the cell to split in half!

Spindle: A structure that reaches out to grab and pull the two chomosomes apart. It is like long fingers that reach out to grab and pull the chromosomes apart once they are ready to split.

Note to parents:

Parents, guardians, and storytellers alike,
This book is a simplified version of the complex process of mitosis. There are other components and structures that are omitted in order to better help your child grasp the core concept of mitosis. "Happy Cells!" is written to provide your child the building blocks of this important process in a fun, creative, and inspiring way, preparing them as they move forward in learning about the amazing human body. Below you will find web sites and videos that explain the terms covered in this book. We encourage you to read, watch and explain them to your child as they read along! Happy learning!

https://www.youtube.com/watch?v=L0k-enzoeOM
https://www.youtube.com/watch?v=RNwJbMovnVQ
https://www.yourgenome.org/facts/what-is-mitosis
https://www.khanacademy.org/science/biology/cellular-molecular-biology/mitosis/a/phases-of-mitosis

33

Bronwyn Tollefson is a student at the University of St Thomas. Her current studies are in biology and Spanish, and she is involved in undergraduate research. She is also a Collaborative Inquiry grant recipient, studying biological growth and development under Dr. Afshan Ismat. In her free time, Bronwyn enjoys creative writing and is involved in several campus organizations. This has led to her passion of combining the field of biology and creative writing in a fun, inspiring way.

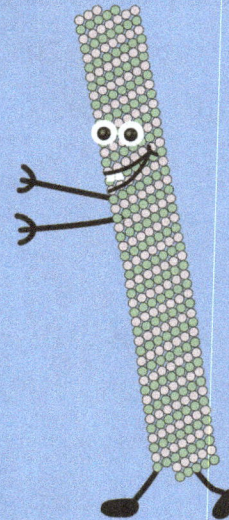

Haude Levesque is a fish biologist who grew up in France. She did her graduate studies in Canada and earned her PhD at the University of Ottawa. She did her post doc studies at the University of Minnesota and is now working as an adjunct professor in the Biology Department of the University of St. Thomas. Haude is also a scientific illustrator since 2008 and has illustrated several science books for children and adults and is the author and illustrator of "Fish Tricks: the wild and wacky world of fish" published by MoonDance Press in 2016.

www.ingramcontent.com/pod-product-compliance
Lightning Source LLC
Chambersburg PA
CBHW082100210326

41521CB00032B/2574